An Introduction to Food Sciences

Engr. Numan Tariq

Engr. M. Waseem Akbar

Copyright © 2012 Nutritional Guardian

All rights reserved.

ISBN: 9798589252323

DEDICATION

This E-Book is dedicated to all Food Engineers and Food Technologist who have worked hard to improve the whole supply chain network of food from farm to fork. Food Engineers and Technologists are working for the previous many years to improve different aspects of food commodities. There is a great role for food professionals in food preservation, food storage, food manufacturing, and food innovation. Food professionals throughout the world have introduced many new innovative techniques to change raw food products into finished ones. So, this book is a dedication to all food professionals around the globe.

.

CONTENTS

ACKNOWLEDGMENTS

This E-book is written by a Food Engineering researcher named Nouman Tariq from the University of Agriculture, Faisalabad, Pakistan. Special thanks and dedications are a tribute to the efforts of Muhammad Waseem Akbar (Food Engineer) who contributed his theoretical knowledge to increase the effectiveness of this book in a very precise manner. Also, a special thanks to the research articles and research papers of senior food technologists and food engineers around the globe who made the completion of this book possible.

1 INTRODUCTION

Food science is a basic and applied science linked with food. It has a deep connection with agriculture sciences, food technology, food safety, and food processing. The IFS defines food sciences as "A branch of applied sciences that uses different engineering, physical, chemical, and biological attributes to study the nature of food, factors that affect deterioration, improvement of food products for public health, and food processing principles".

- Food sciences bring different disciplines together like physics, chemistry, microbiology, and biotechnology

- Food sciences are interlinked with different other fields like Food Science and Technology, Chemical Engineering, Food Process Engineering, and vicinities

1.1 The Father of Food Sciences

Frenchman Nicholas Appert is considered to be the father of food sciences. In the early 1800s, he developed the methods of preservation of food in glass jars. Many food scientists consider Nicholas Appert as the father of food sciences. By the time, food sciences got many more advancements. Modern food sciences involve many more things other than food preservation.

Food science is viewed as an applied science; like engineering, it utilizes information from a few regular logical fields to tackle commonsense issues. Food researchers are knowledgeable about the standards of microbiology, science, and material science. Since food safety is the essential worry of everybody, food researchers work to guarantee that items are not tainted with organisms or destructive substances.

- The synthetic idea of nourishments especially is significant in controlling the nature of flavour, shading, appearance, and surface.

- Food researchers likewise need to have information on engineering (Food and Chemical) standards so they can see how a handling innovation impacts the food.

1.2 Subdisciplines of Food Sciences

There are many interlinked disciplines in food sciences. It covers the study of different scientific fields like chemistry, physics, microbiology, and biotechnology. It is a diversified field that is also linked with engineering troubleshooting. This field is linked with Food Process Engineering as well as Chemical Engineering. Some most important fields that are linked with food sciences and are called as the subdisciplines of food sciences are as follow:

1.2.1 Food Chemistry

Food Chemistry is the study of the different nutritional attributes of food like proteins, carbohydrates, vitamins, fats, lipids, and water.

1.2.2 Food Physical Chemistry

Food Physical Chemistry is the study of both physical and chemical interactions in foods in terms of physical and chemical principles applied to food systems, as well as the application of physicochemical techniques and instrumentation for the study and analysis of foods.

1.2.3 Food Engineering

Food Engineering is linked with the study of the industrial process to manufacture different food commodities.

1.2.4 Food Microbiology

Food Microbiology is the study of different micro-organisms that are linked with food.

1.2.5 Food Technology

Food Technology is linked with the technological and engineering aspects through which we can process food commodities.

1.2.6 Foodomics

Foodomics is characterized as "an order that reviews the Food and Nutrition areas through the application and reconciliation of cutting edge - omics advances to improve shopper's prosperity, wellbeing, and information". Foodomics requires a mix of food science, organic sciences, and information examination.

1.2.7 Molecular Gastronomy

Molecular Gastronomy is linked with the study of the change in physical and chemical properties of food components during cooking.

1.2.8 Quality Control

Quality Control is defined as a science of performing sensory and objective analysis to ensure product safety to avoid food-borne illnesses.

1.2.9 Sensory Analysis

Sensory Analysis deals with the investigation of the different attributes of food commodities through human senses.

1.3 Why there is a need for Food Sciences?

According to the statistical analysis of FAO, It is estimated that about 1.3 billion tonnes of food are wasted per annum globally. Food scientists are striving very hard from the previous many decades to overcome food losses. The food scientists are working to improve both food safety with the premises of food industries and food security. From the previous many decades, food scientists have contributed massive scientific assistance to food industries about the preservation, processing, packaging, storage, and distribution of food. For the amelioration of food security, food scientists are working with the agricultural supply chain experts to mitigate the bad pre-and post-harvest practices.

1.4 What is Food Safety and Security?

There is a difference between food safety and food security. The amelioration of both food safety as well as security is important to ensure an effective food supply chain. Figure 1.1 is depicting the main differences between food safety and security. Food sciences are linked with all of the following aspects of food safety and security mentioned in Figure 1.1.

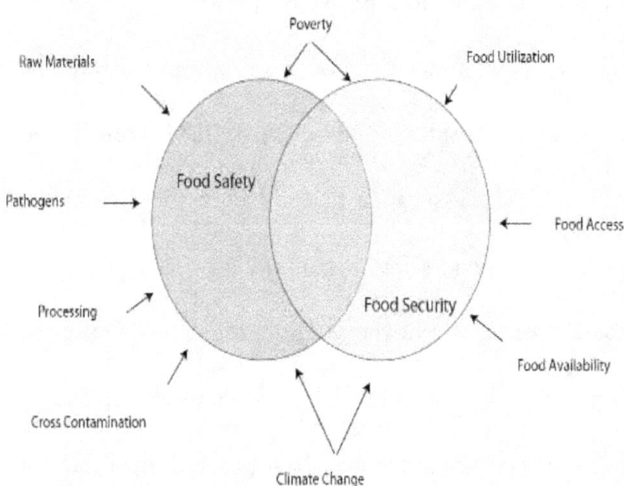

Figure 1.5: *Food Safety and Food Security*

1.5 Development in Food Sciences

Food sciences start from preservation when Nicholas Appert discovered food preservation in glass jars. By the time, Food scientists expand this field and currently, there is a very broader spectrum of food sciences. Now, food sciences include the following aspects:

1.5.1 Aspects of Modern Food sciences

- Food preservation

- Food Research and development

- Food safety and security

- Food processing

- Food quality control and quality assurance

- Food analysis (Physical, Chemical, Technical, and Biological)

- Foodomics

- Food engineering

- Sensory and Objective analysis of food products

- Food chemistry and biochemistry

1.5.2 Scope of Food Sciences

Food scientists can perform multiple roles to ensure a good food supply chain. They can work as food industries for their quality and production department. Food scientists can also contribute effectively to the R&D department of the food industry. Food auditing is an emerging field from the previous few years. Food Scientists can also serve auditing companies for the audit of different food industries. The scope of food sciences is on a very broad spectrum and its importance is increasing day in and day out.

Dreams are not that you see while sleeping, it is something
that does not let you sleep.

-- A.P.J Abdul Kalam Azad

2 FOOD CONSTITUENTS

The Food that we eat consists of six basic nutrients. These nutrients are the part of almost every food commodity and their intake is very necessary for effective mitigation of different diseases. These nutrients are:

1.   Carbohydrates

2.   Proteins

3.   Lipids/ Fats

4.   Vitamins

5.   Minerals

6.   Water

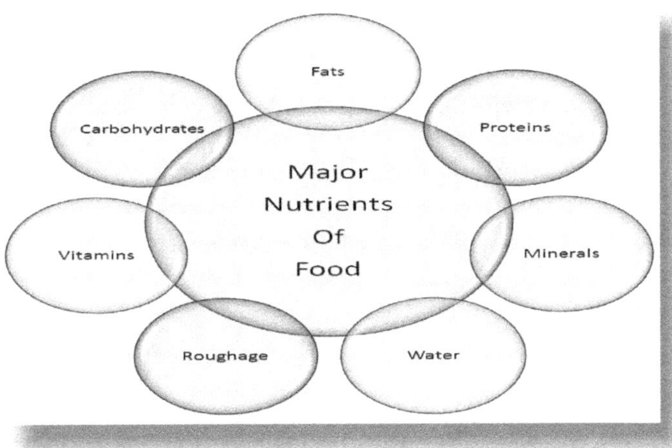

Food commodities are classified into various classes according to the percentage composition of these mentioned nutrients. The purpose of this chapter is to depict the introductory information about these nutrients.

2.1 Carbohydrates

Carbohydrates are the biomolecules that are made up of carbon, hydrogen, and oxygen. Their general formula is generally given as $C_n(H_2O)_n$. The value of m and n in the formula varies according to different scenarios. Carbohydrates are the hydrates of carbon and technically they can be viewed as aldoses and ketoses.

The term carbohydrates are more related to biochemistry and they have a synonym which is a saccharide. The saccharide is classified into four major groups.

2.2 Groups of Saccharides

The saccharides are classified as:

- Monosaccharide

- Disaccharide

- Oligosaccharides

- Polysaccharides

Monosaccharides and disaccharides, the littlest (lower sub-atomic weight) Carbohydrates that are regularly alluded to as sugars. The word saccharide originates from the Greek word sákkharon, signifying "sugar". While the logical classification of sugars is unpredictable. lactose sugar is a disaccharide and found in milk. The structural formula of lactose is depicted in Figure 2.1.

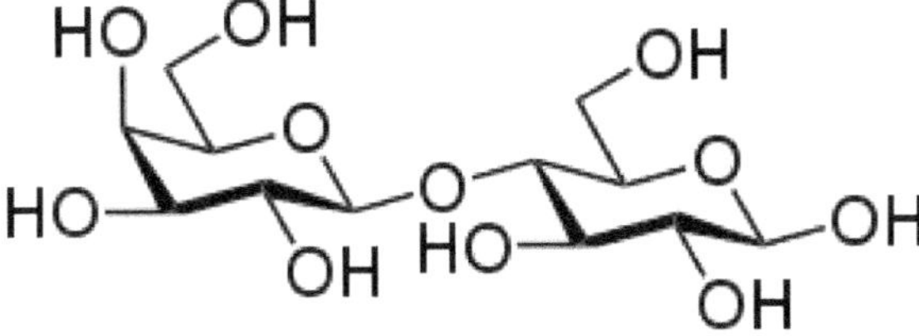

Figure 2.1: *Lactose structural formula*

2.3 Monosaccharides

These are referred to be as carbohydrates that are simplest in nature. Their main types are Fructose, glucose, galactose, and mannose. These are the simplest sugars and they do not hydrolyze further. They have the general formula $C_nH_{2n}O_n$. There are generally colourless, and water-soluble. They are sweeter in taste.

Monosaccharides are the building blocks of all other groups of carbohydrates. When two monosaccharides combine, they form disaccharides like sucrose, maltose, and lactose (Milk Sugar). They are also the building blocks of polysaccharides such as starch, pectins, and gums that are highly complex glucose long-chain polymers.

2.3.1 Structure of Some Monosaccharides

The structure of glucose, fructose, mannose, and galactose are as follow:

(i)   Structure of Glucose

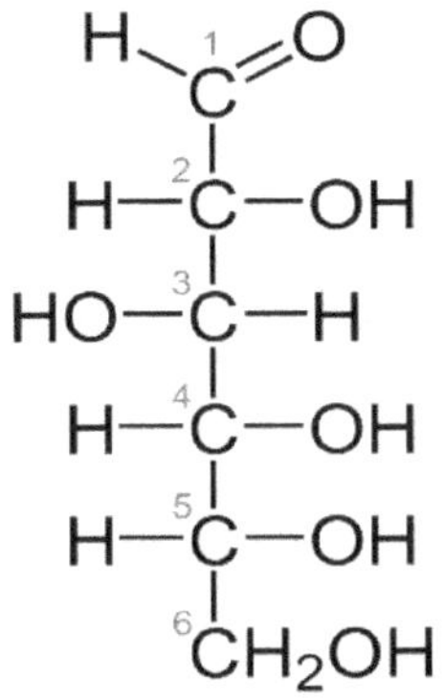

$$\begin{array}{c} H-\overset{1}{C}=O \\ | \\ H-\overset{2}{C}-OH \\ | \\ HO-\overset{3}{C}-H \\ | \\ H-\overset{4}{C}-OH \\ | \\ H-\overset{5}{C}-OH \\ | \\ \overset{6}{C}H_2OH \end{array}$$

Figure 2.2: *Structure of Glucose*

(ii)   Structure of Fructose

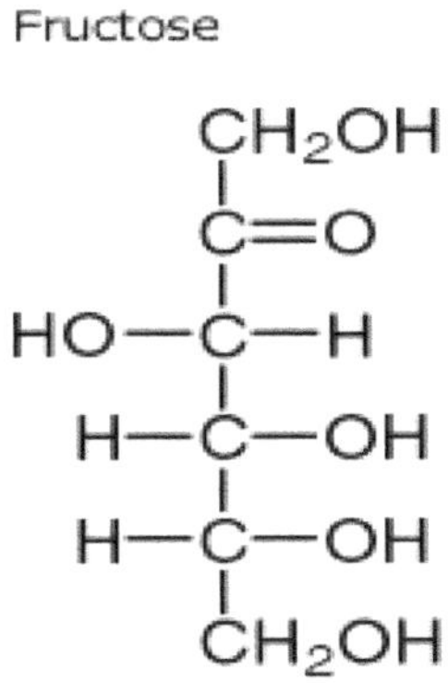

Fructose

$$\begin{array}{c} CH_2OH \\ | \\ C=O \\ | \\ HO-C-H \\ | \\ H-C-OH \\ | \\ H-C-OH \\ | \\ CH_2OH \end{array}$$

Figure 2.3: *Structure of Fructose*

(iii)   Structure of Mannose

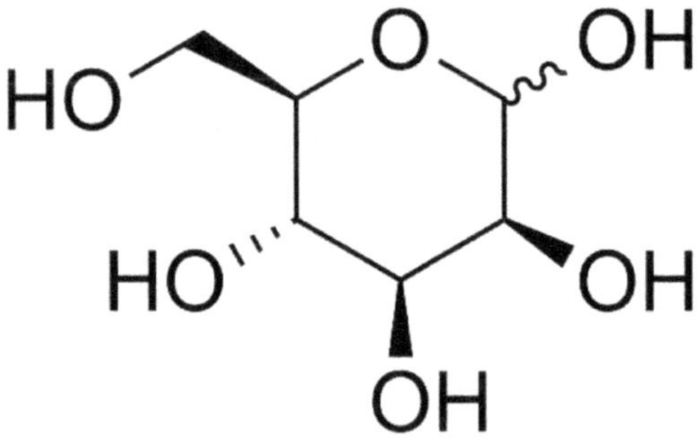

Figure 2.4: *Structure of Mannose*

(iv)   Structure of Galactose

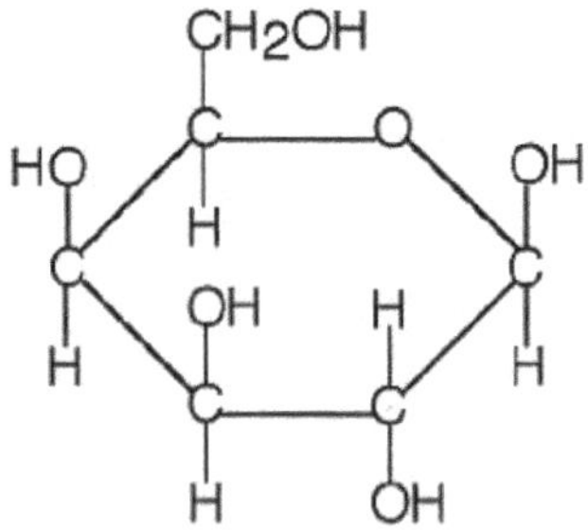

Figure 2.5: *Structure of Galactose*

2.4 Disaccharides

Disaccharides are also carbohydrates that form by the combination of two monosaccharides. They are also known to be as double sugar. These are the carbohydrates that are soluble in water. There are three main types of disaccharides. These disaccharides are the combination of two monosaccharides.

2.4.1 Examples of Disaccharides

The three examples of disaccharides are as follow:

- Sucrose (Table Sugar)

- Maltose (Malt Sugar)

- Lactose (Milk Sugar)

2.4.2 Formation of Disaccharides

The monosaccharides combine to form disaccharides with the elimination of water. Monosaccharides undergo condensation reactions. A condensation reaction is one in which the water is eliminated. The reaction between two monosaccharides to form disaccharides is a dehydration synthesis reaction in which water is produced as a by-product when –OH group from both monosaccharides combine.

2.4.3 Simple Reactions

1)   Glucose + Glucose   → Maltose + Water

2)   Glucose + Fructose   → Sucrose + Water

3)   Glucose + Galactose   → Lactose + Water

2.4.4 Complex Structural Reactions

(i) Formation of Sucrose

Here in the following reaction 6-membered glucose molecule is combined with the 5-membered fructose molecule which has a ketone group rather than the aldehyde group. We know that sucrose is the disaccharide of glucose and fructose. This is common table sugar and it comes from sugar cane and sugar beets. Maple syrup also contains sucrose. We are well aware of the fact that the reaction between the two monosaccharides molecules undergo dehydration synthesis and eliminate water due to a combination of –OH group from both sugars.

The complex structural reaction for the formation of sucrose is as follow:

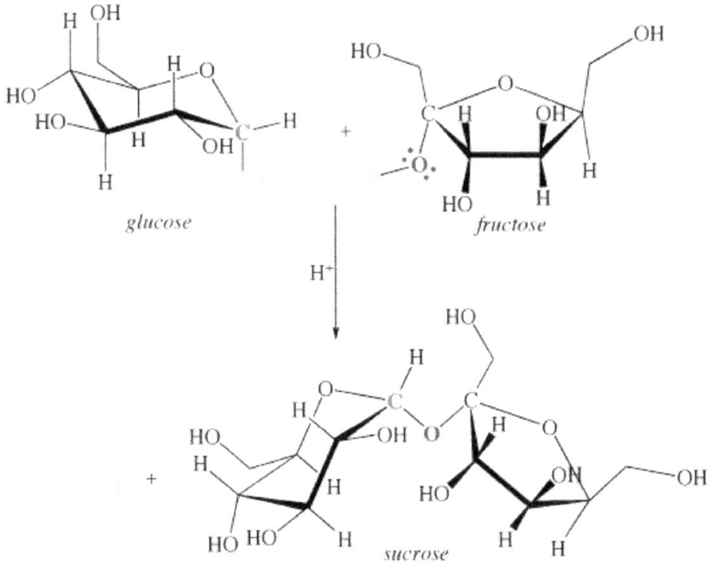

Figure 2.6: *Reaction to the formation of sucrose*

(ii) Formation of Maltose

Maltose is a disaccharide form by the combination of two glucose molecules. The complex structure of the reaction follows:

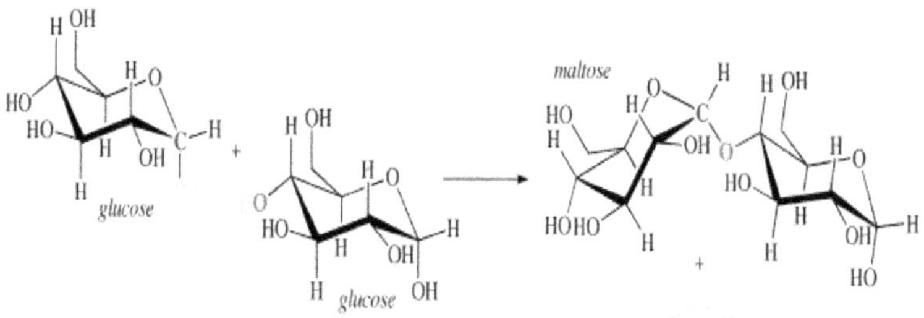

Figure 2.7: *Reaction to the formation of Maltose*

2.5 Oligosaccharides

Oligosaccharides are the carbohydrates that consist of 3-10 sugar units joined together. There are many functions of oligosaccharides including the cell recognization and cell binding. They are made of any sugar monomers.

We can say that oligosaccharides having three sugar units

are called as trisaccharides. There are many famous examples of trisaccharides such as Raffinose and Stachyose. Raffinose is the combination of glucose, fructose, and galactose while Stachyose is the combination of glucose, fructose, and two galactose units.

2.6 Polysaccharides

Polysaccharides are the carbohydrates that consist of many long chain glucose polymers. There are many best examples of polysaccharides such as starch, gums, and pectins. Starch has very extensive applications in food industries, especially in the bakery and confectionery industries.

It is important to mention that starch is a long chain polymer which under the process of gelatinization.

2.7 What is Gelatanization?

Gelatinization is a process in which the starch is mixed

with the water and shows the thickness and gelling properties by providing temperature. It should be noted that the process of gelatinization is a TCS (Time-Temperature Control System) and it is always challenging to undergo this process efficiently.

2.8 Vitamins

Vitamins are the organic compounds that are referred to be as micro-nutrients that are required for the proper functionalities of metabolism. They are included in the list of essential nutrients and many times cannot be synthesized by the body.

2.8.1 Classification of Vitamins

Vitamins are classified as water-soluble and fat-soluble. Our body consists of a total of 13 vitamins. 4 fat-soluble vitamins in our body are Vitamin A, D, E, K, and 9 water-soluble vitamins which include 8 B-Vitamins and Vitamin C which is also known as ascorbic acid.

Water-soluble vitamins dissolve easily in water and, in general, are readily excreted from the body to the degree that urinary output is a strong predictor of vitamin consumption. Because they are not as readily stored, more consistent intake is important. Fat-soluble vitamins are absorbed through the intestinal tract with the help of lipids (fats). Vitamins A and D can accumulate in the body, which can result in dangerous hypervitaminosis. Fat-soluble vitamin deficiency due to malabsorption is of particular significance in cystic_fibrosis.

Note

It should be noted that the intake of vitamins should be in a controlled manner. We should not intake too many fat-soluble vitamins like Vitamin A, D, E, and K. The excessive intake of fat-soluble vitamins can cause many problems, especially Hypervitaminosis. It is a problem due to excessive intake of Vitamin A. It can cause many problems like vision issues, skin changes, and bone pain.

(i)   Water-Soluble Vitamins

The following vitamins are included in water-soluble vitamins:

- Thiamine(B-1)

- Riboflavin(B-2)

- Niacin(B-3)

- Pantothenic acid(B-5)

- Pyridoxine(B-6)

- Biotin(B-7)

- Folate(B-9)

- Cobalamin(B-12)

- Ascorbic acid

(ii)   Fat-Soluble Vitamins

The following vitamins are included in fat-soluble vitamins:

- Vitamin A

- Vitamin D

- Vitamin E

- Vitamin K

2.9 Proteins

Proteins are the class of organic compounds or more precisely any class of nitrogenous organic compounds that consist of one or more long chain of amino acids. They are very important for different functionalities of the body. They are of primary importance for the body tissues like muscles and hair. There are different types of proteins and their types depend upon the sequence of amino acids. Whereas, the sequence of amino acids is linked with the nucleotide sequence of their genes. The nucleotide genes sequences make 3-D structures of a protein that determines its activity.

2.9.1 3-D Structure of Protein Myoglobin

The following structure (Figure 2.8) is the structure of myoglobin protein. Here we can see how the sequence of amino acid and nucleotide gene affects the structure of the whole protein. The myoglobin is the first protein in which structure was solved by X-ray Crystallography.

In Figure 2.8, we can see that there is a central prosthetic group called a Heme group (Grey in Color) with oxygen bounded in it (Red in Color).

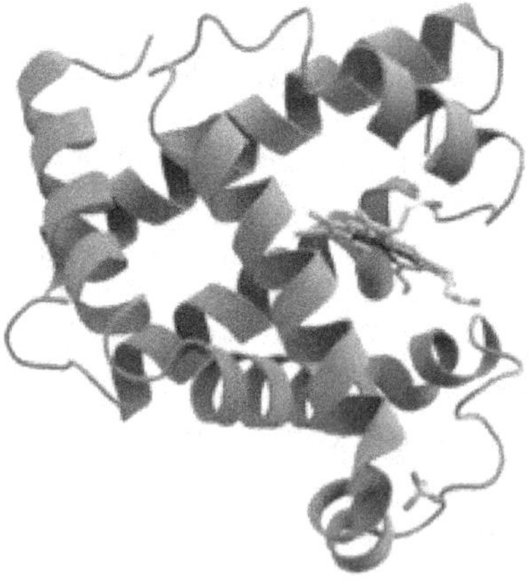

Figure 2.8: *3-D Structure of Myoglobin*

The protein content of animal organs is usually much higher than that of the blood <u>plasma</u>. Muscles, for example, contain about 30 per cent protein, the liver 20 to 30 per cent, and <u>red blood cells</u> 30 per cent. Higher percentages of protein are found in hair, bones, and other organs and tissues with low <u>water</u> content. The quantity of free amino acids and <u>peptides</u> in animals is much smaller than the amount of protein; protein molecules are produced in <u>cells</u> by the stepwise alignment of amino acids and are released into the body fluids only after synthesis is complete.

2.9.2 Haemoglobin- A protein

In all vertebrates, the respiratory protein <u>haemoglobin</u> acts as an <u>oxygen</u> carrier in the <u>blood</u>, transporting oxygen from the <u>lung</u> to body organs and tissues. A large group of structural proteins maintains and protects the structure of the animal body.

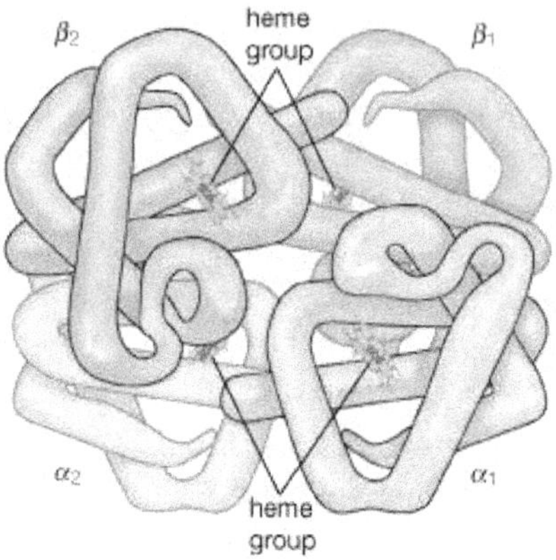

Figure 2.8: *Structure of Hemoglobin*

Figure 2.8 is depicting that the molecule of haemoglobin is made up of four peptide chains named as α_1, α_2, β_1, and β_2. Each chain is attached to a heme group composed of porphyrin (an organic ringlike compound) attached to an iron atom. These iron-porphyrin complexes coordinate oxygen molecules reversibly, an ability directly related to the role of haemoglobin in oxygen transport in the blood.

2.10 Minerals

Minerals are inorganic substances that are required by the

body to perform different functions. They are involved in the effective formation of bones and teeth. There is also a part of complex enzyme systems and also plays an important role in the control of the nervous system.

They are classified as Macro and Micro Minerals. The names of Micro and Macrominerals are as follow:

2.10.1 Micro-Minerals

- Iron

- Zinc

- Selenium

- Iodine

- Copper

- Manganese

- Fluoride

- Chromium

2.10.2 Macro-Minerals

- Sodium

- Calcium

- Phosphorous

- Potassium

- Magnesium

- Sulfur

2.11 Water

Water is also a basic nutrient that is required for the effective control of body functions. It is a colourless, transparent, odourless liquid that forms the seas, lakes, rivers, and rain and is the basis of the fluids of living organisms.

2.11.1 Nutritional Definition of Water

Water is defined as an essential nutrient because it is required in amounts that exceed the body's ability to produce it. All biochemical reactions occur in water. It fills the spaces in and between cells and helps form structures of large molecules such as protein and glycogen.

2.11.2 Why water is important?

- Your body uses water in all its cells, organs, and tissues to help regulate its temperature and maintain other bodily functions.

- Because your body loses water through breathing, sweating, and digestion, it's important to rehydrate by drinking fluids and eating foods that contain water.

2.11.3 Food Products Classification according to Water content

Water also plays an important role in different food products. Knowing the chemistry of water, its role in food products and its effects are a major study area of food sciences. The food that we eat is generally divided into three types according to the water percentage. These classifications are as follow:

- Perishable commodities

- Semi-Perishable commodities

- Non-Perishable commodities

Knowing all the mentioned classifications is a major area of study for food scientists.

Development is being sought in every walk of life and you have to take on this process of development. Are you preparing to take on tomorrow's responsibilities? Are you building your capacity? Are you trained enough? If no, then go and prepare yourself because this is the time to prepare yourself for future responsibilities.

-- Muhammad Ali Jinnah (Quaid-e-Azam)

3 CLASSIFICATION OF FOOD COMMODITIES

The study of food product classification is very extensive and interesting. The food that we eat is classified according to different categories. The food commodities are may be classified into fresh organic and processed commodities. Some categorization of food commodities is as follow:

- According to water content (Aw values)

- According to nutritional values

- According to shelf-life

- According to nature (Organic or processed)

- According to chemical properties (Acidity, Alkalinity, and PH values)

- According to rheological attributes

- According to Maturity (Measuring Maturity Index)

Here in this chapter, we will discuss the food commodities according to water content and chemical properties.

3.1 Classification According to Water Content

(i) Moisture Content

Moisture content is the quantity of water which is present in food product either free-form, chemical bonded, and physically bonded form.

We generally find out the % moisture contents of food products during drying and vicinity processes.

(ii) Water Activity

Water activity is defined as the availability of water to take part in different types of chemical and biological reactions.

Being a Food scientist, we should remember that water activity has more concern with the free-water in food products. It has no link with chemically bonded and physically bonded water.

It is represented as Aw and its maximum value is always 1.

FDA (Food and Drug Administration) has maximized the Aw value of many food products as 0.85.

The food commodities are classified into three main types according to water percentages. Water plays a constructive as well as a destructive role for different food products. The three types are as follow:

- Perishable commodities

- Semi-Perishable commodities

- Non-Perishable commodities

All mentioned commodities classification is linked with water percentage. The water content in food products is linked with the % moisture content and water activity (A_w).

3.1.1 Perishable Commodities

Perishable commodities are commodities that have a high level of water content. These products have a high value of water activity and can deteriorate rapidly.

Figure 3.1: *Different Perishable Food Commodities*

(i) Dairy Products and Eggs

Dairy products are those commodities, which are derived from or based upon milk, and include creams, yoghurts, butter, cheese, and ice cream. Milk needs to be stored in the refrigerator at a temperature between 3 to 4°C. If stored this way fresh milk will last about 10 days. Cream, yoghurts, butter, cheese, and eggs should also be stored at between 3 to 4°C. The shelf life of these products will vary depending on their method of manufacturing, and you should check individual use by dates on the packaging. Ice cream needs to be kept frozen at a temperature of -18°C or below. All dairy products need to be kept well sealed when in storage, otherwise they will absorb flavours from strong-smelling foods around them.

(ii)   Meat and Poultry

Meat and poultry should be stored between 1 and 3°C. All meat and poultry should be stored on clean trays and covered with plastic wrap. You should never store raw and cooked foods on the same tray.

Meat can be stored in the cool room for 4-6 days, or if vacuum-sealed up to 12 weeks. Poultry can be kept for 3-4 days. If frozen, meat and poultry can be kept for up to six months.

(iii)   Seafood

Seafood has a very short shelf life and must be stored with extra care. It should be stored at 1°C. Unfortunately, most cool rooms are not set for temperatures as low as this, so seafood should be wrapped in plastic film and stored in the coldest part of the cool room on a bed of crushed ice. If kept in these conditions' seafood should last for 5 to 6 days.

Live seafood such as crabs and yabbies' should be kept at temperatures between 1 to 3°C in sealed containers. Frozen seafood can be kept in the freezer for up to 3 months.

(iv)   Cooked Foods and Leftovers

These are also considered perishable and should be stored in the cool room at between 3 to 4°C. Cooked foods should be covered before storage and need to be stored separate from raw foods and never on the same tray. Always allow hot foods to completely cool before refrigeration.

(v)   Fruit and Vegetables

Fruit and vegetables vary in their storage requirements, but as a general rule, most fruit and vegetables should be stored between 5 and 9°C. There are a couple of exceptions such as broccoli, which usually arrive packed on ice, and should be stored at 1°C, and tropical fruits such as bananas and pineapples should be stored at around 18°C.

Because fruit and vegetables require higher temperature storage (between 7 and 10°C) they are best kept in a separate cool room.

Lettuce should be stored in the cool room, where the temperature is below 5°C.

Root vegetables, such as carrots, potatoes, and onions are classed as semi-perishable and do not require refrigeration.

Frozen vegetables and fruit are stored in a deep freezer where the temperature is set at -18°C or less.

Processed vegetables and fruit come in cans, jars, and packages. They should be stored on shelves in a cool dry room away from sunlight. It is important to rotate this stock.

3.1.2 Semi-Perishable Commodities

Semi-perishable commodities are those that do not require refrigeration but still have a limited shelf life. They include things like potatoes, onions, pumpkins, and salamis. These items are usually kept on shelves in the storeroom complex, where they get plenty of air circulation around them. Potatoes need to be kept away from light as they will start sprouting.

Figure 3.2: *Different Semi-Perishable Commodities*

3.1.3 Non-perishable Commodities

Technically speaking there is no such thing as non-perishable commodities, as all goods deteriorate overtime. But some commodities deteriorate so slowly that they are called non-perishable. Examples of non-perishable goods are:

- Flour

- Spices

- Canned foods

- Jars and bottles

- Nuts

- Dried packet goods, for example, noodles and pasta

These items are usually kept in the dry store where they are kept cool and are protected from moisture contamination

Figure 3.3: *Different Non-Perishable Commodities*

3.2 Classification According to Chemical Properties

There are different chemical and nutritional properties of food products. Food products can be classified according to their acidic and basic character. We know that the neutralize PH range is 7. The food Products having PH less than 7 (<7) are acidic in nature and food products having PH >7 is basic in nature.

Education is the most powerful weapon which you can use
to change the world. For to be free is not merely to cast
off one's chains, but to live in a way that respects and
enhances the freedom of others

--Nelson Mandela

4 FOOD SPOILING AGENTS

Food that we eat is divided into different categories. Each category has its specification and resistance towards the spoiling agents. Firstly, we will discuss what are food spoiling agents and into how many types they are classified.

4.1 Food Spoiling Agents

Food spoiling agents are agents that change the physical, chemical, and rheological properties of different food commodities. They decrease the shelf-life of food products by changing their texture and other factors like colour, aroma, and taste. They are divided into different categories.

4.2 Categorization of Food Spoilage Agents

The main classification of food spoilage agents is as follow:

1.   Micro-organisms

2.   Enzymes

3.   Birds, rodents, and insects

4.   Physical factors

5.   Chemical Factors

4.2.1 Micro-organisms

There are many micro-organisms like bacteria, viruses, mould, yeast, and fungi. These micro-organisms cause food spoilage in different ways.

(i)   Bacteria and Food Spoilage

- Lactic acid formation: Lactobacillus, Leuconostoc

- Lipolysis: Pseudomonas, Alcaligenes, Serratia, Micrococcus

- Pigment formation: Flavobacterium, Serratia, Micrococcus

- Gas formation: Leuconostoc, Lactobacillus, Proteus

- Slime or rope formation: Enterobacter, Streptococcus

(ii)   Moulds and Food Spoilage

Some strains produce mycotoxins under certain conditions. Mycotoxins can penetrate the parts of food that are not visibly mouldy as well. It is, therefore, necessary to throw away all of the food if any part of it is mouldy. They are also notoriously difficult to destroy as they are stable to both heat and chemicals.

(iii)   Foods most at Risk for Molds Formation

1.   Grains and grain products - many mycotoxin types

2.   Peanuts, nuts and pulses - aflatoxin

3.   Fruits and vegetables (raw and preserved) - patulin

4.   Milk and milk products - aflatoxin should be noted that if the animal's feed is contaminated with any

strain of mould then the products obtained from animals also contain harmful micro-organisms like Aflatoxin.

(iv) Pathogenic Food Micro-Organisms

Pathogenic micro-organisms cause food-borne infections or intoxication and include bacteria, viruses, parasites, and moulds. It is important to note that pathogenic bacteria and viruses usually do not cause food spoilage, their contamination cannot be seen nor tasted.

The main factors that contribute to the occurrence of foodborne diseases are:

1.   The use of raw food and ingredients from unsafe sources

2.   Inadequate cooking or heat processing

3.   Improper cooling and storing, for example leaving cooked foods at room temperature for longer

periods, or storing foods in large containers in the fridge

4.   Allowing several hours to pass between preparation and eating of food

5.   Inadequate reheating

6.   Improper hot holding, meaning below 65°C

7.   Food handling by infected persons or carriers of infection

8.   Cross-contamination from raw to cooked food. For example, cutting vegetables for salad on a cutting board where you have cut raw meat before

9.   Inadequate cleaning of equipment and utensils

4.2.2 Physical Factors

There are different physical factors in food industries that cause food spoilage during different industrial operations like processing, storage, packaging, and transport.

The main physical factors that affect food spoilage are as follow:

- Glass and plastic pieces

- Metallic pieces

- Wooden residues

- Weeds, stones, and unwanted particles

These factors can affect the whole food supply chain network. Food safety is, therefore, is a major concern to mitigate food spoilage.

(i) Other Physical Factors

Many other physical factors can cause food spoilage. Some are discussed below:

- Temperature

- Moisture

- Pressure

- Heat

Moisture and heat control is of great importance in oils and fats. The bad control of moisture or the entrance of oxygen into the fats and oils causes the fats to turn into fatty acids. These fatty acids produce a rancid odour and cause oil rancidity

Excessive temperature can also affect food products. The bad management of TCS (Time-Temperature Control System) can cause the denaturation of proteins, breakage of emulsions, and destruction of nutrients like vitamins.

Pressure can also spoil the food and change its shape bt changing its rheological properties.

4.2.3 Chemical Factors

Chemical food spoilage is the production of different changes in the food products like staling, discolouration, and formation of off-odours (Rancidity).

- Texture – such as the thinning of sauces from starch degradation.

- Colour – including discolouration of meat, fruit, or vegetables.

- Off-odour and Off-flavour:

- Breakdown of proteins in meat.

- Taint from packaging, environment, or processing.

- Rancidity from lipid oxidation or hydrolysis.

4.2.4 Birds, Rodents, and Insects

The processing of food products in food industries is a great challenge for food scientists. The conversion of raw material into safe and sound food products is due to the consecutive efforts of Food Technologists and Food Engineers. Food premises especially production areas should be free from the entrance of different birds, rodents, and insects.

The rodents and insects are very notorious for the food spoilage as they can entirely change the physical and chemical properties of food products.

Food industries especially, MNCs follow proper food safety techniques to implement IPM (Integrated Pest Management) to mitigate the rodents, insects, and vicinity species from the industry.

4.3 Enzymes and Food Spoilage

The storage of foods is limited by nonenzymatic, enzymatic, or microbial reactions that alter the edible quality of foods, including deterioration, appearance, texture, aroma, flavour, nutrition, and safety and functional properties. Enzymatic reactions with respiration at the tissue surface can lower oxygen concentration and indirectly promote non-enzymatic oxidation of myoglobin. When the raw milk is stored for a long time at low temperature before heat treatment, the psychrotrophic bacteria can multiply and produce extracellular heat-stable enzymes. Proteinases and lipases produced by psychrotrophic bacteria in raw milk can cause noticeable hydrolysis of proteins and lipids within 3–7 days. The heat-stable enzymes produced by psychrotrophic bacteria can cause spoilage of food products. Cream and butter are more susceptible to spoilage by heat-stable lipases than proteinases. The bacterial proteinases and lipases from psychrotrophic bacteria such as *Pseudomonas*, *Aeromonas* cause flavour and texture defects in raw meat and fish.

People have become educated, but have not yet become human.

--Abdul Sattar Edhi (Philanthropist)

5 FOOD PRESERVATION

Food preservation is not a new thing. People started it at an early age. It dates back to the 1700s when a French baker responded to a call to develop a new method to preserve food for the army. His method was widely accepted and quickly used all over. By the 1900s canning made its way to the U.S. and remains the most popular method of food preservation today.

There are many methods through which we can preserve food products. Some important food preservation methods are as follow:

- High-temperature technique

- Low-temperature technique

- Process of irradiation

- Use of chemical additives

- Process of fermentation

- Food Packaging

5.1 High-Temperature Techniques

There are many high-temperature techniques through which we can control the food spoilage and promote food preservation.

Some of the high-temperature techniques are as follow:

- Cooking

- Blanching

- Pasteurization

- Canning

5.1.1 Cooking

Cooking is applied to different food products to change its colour, flavour, and textural properties. Food is cooked to change its texture and digestibility. The cooked food products have a longer shelf-life as compared to the uncooked food products. There are different ways of cooking. We can cook food products by direct heat exposure and also by the dipping of food products in a suitable amount of water.

Cooking embraces different types of heating methods like roasting, boiling, broiling, frying, stewing, and steaming.

5.1.2 Blanching

Blanching is a hot water treatment process in which is employed to food products before the process of freezing, canning, cold storage, and dehydration. Blanching reduces the enzymatic browning of food commodities like potatoes. We undergo the process of blanching for the apples, potatoes, and vicinity products to restrict the change of colour.

5.1.3 Pasteurization

Pasteurization is also a process of preservation. It is generally employed for liquid food products. Pasteurization can be continuous or batch. The continuous pasteurization process includes LTLT and HTST process. There is always a time-temperature control system in the pasteurization process.

LTLT involves heating the milk to 62.5°C/144.5°F and holding this for 30 minutes and is the method used by milk banks that perform either the Holder method of pasteurization or the similar Vat method of pasteurization.

HTST pasteurization uses stainless steel heat exchange plates where the product flows on one side while the heating media flows on the opposite side to raise milk temperatures to at least 161° F (72°C) for at least 15 seconds*, followed by rapid cooling. It is also known as the flash pasteurization process.

5.1.4 Canning

Canning is also a preservation process. In this process, food is hermetically packed in containers and dipped in a suitable amount of salt or sugar solution. It is a process through which we can increase the shelf life of different organic and fresh food products like fruits and vegetables. Many fruits like apple, pear, and plums are cut with the help of a multi-purpose machine according to desire shape and size. After cutting the mentioned products the products are dipped into the sugar or brine solution.

Note

It is very important to note that the formation of the sugar solution is very important. The sugar for the formation of sugar syrup should be refined. The unrefined sugar consists of hydrogen which produces hydrogen sulfide which creates the black deposits on the can body.

5.2 Low-Temperature Technique

Some of the low-temperature techniques for food preservation are as follow:

- Refrigeration

- Freezing

- Cold storage technique

Refrigeration and freezing are techniques used for preserving food in low temperatures. These procedures slow down or stop most bacteria from dividing and thereby multiplying, but do not kill them.

Many micro-organisms are spore-forming like Bacilli, Clostridium Botulinum, and Clostridium perfringens. These microbes are not able to be killed by low-temperature technique but their growth rate can be decreased.

5.3 Process of Irradiation

Food irradiation (the application of ionizing radiation to food) is a technology that improves the safety and extends the shelf life of foods by reducing or eliminating microorganisms and insects. Like pasteurizing milk and canning fruits and vegetables, irradiation can   make food safer   for   the

consumer.

Food irradiation is the way toward uncovering food and food bundling to ionizing radiation. Ionizing radiation, for example, from gamma beams, x-beams, or electron shafts, is vitality that can be sent without direct contact to the wellspring of the vitality (radiation) equipped for liberating electrons from their nuclear securities (ionization) in the focused-on food. The radiation can be discharged by a radioactive substance or created electrically. This therapy is utilized to improve sanitation by broadening item timeframe of realistic usability (protection), lessening the danger of foodborne ailment, postponing or disposing of growing or maturing, by sanitization of nourishments, and as a method for controlling bugs and obtrusive pests. Food light fundamentally expands the period of usability of illuminated food sources by viably obliterating living beings answerable for deterioration and foodborne sickness and hindering sprouting.

5.4 Use of Chemical Additives

Preservatives are added to food to fight spoilage caused by bacteria, moulds, fungus, and yeast. Preservatives can keep food fresher for longer periods, extending its shelf life. Food preservatives also are used to slow or prevent changes in colour, flavour, or texture and delay rancidity.

Sr no.	E-Number	Chemical Compound	Food Products
1	E200-E203	sorbic acid, sodium sorbate, and sorbates	common for cheese, wine, baked goods, personal care products
2	E210-E213	benzoic acid and benzoates	used in acidic foods such as jams, salad dressing, juices, pickles, carbonated drinks, soy sauce
3	E214-E219	parabens	stable at a broad pH range, personal care products
4	E220-E228	sulfur dioxide and sulfites	common for fruits, wine
5	E249-E50	nitrites	used in meats to prevent botulism toxin

Table 5.1: Important additives with their food applications and E-number

5.5 Food Fermentation

Fermentation is an essential metabolic phenomenon that takes place in the absence of oxygen(O2). In the process of fermentation, sugar is consumed in the absence of oxygen. The products formed due to fermentation are organic acids, gases, or alcohol. Fermentation occurs commonly in yeast and bacteria and also in oxygen-starved muscle cells, as in the case of lactic acid fermentation. In terms of microbiologists, fermentation is a primary means of producing ATP by the degradation of organic nutrients anaerobically, in the presence of suitable microorganisms. Zymology is the science of fermentation. Fermentation processes are believed to have been developed to preserve fruits and vegetables for times of scarcity by preserving the food by organic acid and alcohols, impart desirable flavour, texture to foods, reduce toxicity, and decrease cooking time. Fermentation is a desirable process and the products that are being prepared by natural fermentation are fermented. That includes pickles, sauerkraut, yoghurt, kimchi, cheese, etc.

Education is the passport to the future, for tomorrow
belongs to those who prepare for it today
-- Malcolm X

6 FOOD ADDITIVES

Food Additives are chemicals that alter different properties
of food products. There are different uses of food
additives. They are classified on a very unique spectrum.

They can be used to enhance the shelf-life, stability, rheological properties, elasticity, flavour, aroma, colour, and vicinity. Food additives are substances added to <u>food</u> to preserve <u>flavour</u> or enhance its taste, appearance, or other qualities. Some additives have been used for centuries; for example, preserving food by <u>pickling</u> (with <u>vinegar</u>), <u>salting</u>, as with <u>bacon</u>, preserving <u>sweets</u>, or using <u>sulfur dioxide</u> as with <u>wines</u>. With the advent of processed foods in the second half of the twentieth century, many more additives have been introduced, of both natural and artificial origin. Food additives also include substances that may be introduced to food indirectly (called "indirect additives") in the manufacturing process, through packaging, or during storage or transport.

6.1 Classification of Food Additives

The main classification of food additives are and their uses are as follow:

6.1.1 Acidulants

Acidulants impart sour or acid taste. Common accidents are vinegar, citric acid, tartaric acid, malic acid, fumaric acid, and lactic acid.

6.1.2 Acidity regulators

Acidity regulators are used for controlling the pH of foods for stability or to affect the activity of enzymes.

6.1.3 Anticaking agents

Anticaking agents can control powder products such as milk powder from caking or sticking.

6.1.4 Antifoaming and foaming agents

Antifoaming agents prevent foaming in foods.

6.1.5 Antioxidants

Antioxidants such as vitamin C can prevent oxidation

reactions in food products. Oxidation reactions can change the colour and taste of food products.

6.1.6 Bulking agents

Bulking agents such as starch are additives that increase the bulk of a food without affecting its taste.

6.1.7 Food colouring

<u>Colourings</u> are added to food to replace colours lost during preparation or to make food look more attractive.

6.1.8 Fortifying agents

There are different fortifying agents in food products that increase their nutritional value.

6.1.9 Colour retention agents

In contrast to colourings, colour retention agents are used to preserve the colour of food products. i.e. $K_2S_2O_5$ (Potassium Meta-Bisulphite)

6.1.10 Emulsifiers

Emulsifiers allow water and oils to remain mixed together in an emulsion, as in mayonnaise, ice cream, and homogenized milk. i.e. CMC (Carboxymethyl cellulose).

6.1.11 Flavours

Flavours give food a particular taste or smell and may be derived from natural sources or created artificially.

6.1.12 Flavour enhancers

Flavour enhancers increase the flavour of food. A popular example is monosodium glutamate.

6.1.13 Flour treatment agents

Flour treatment agents are added to flour to improve its colour and rheological properties.

6.1.14 Glazing agents

Glazing agents provide a shiny look or protective coating to foods.

6.1.15 Humectants

Humectants prevent foods from drying. They can control the water activity of food products.

6.1.16 Tracer gas

Tracer gas prevents foods from being exposed to the atmosphere, thus guaranteeing shelf life.

6.1.17 Preservatives

Preservatives prevent or inhibit spoilage of food due to different micro-organisms.

6.1.18 Stabilizers

Stabilizers can control the thickness and gelling properties.

Agar or pectin (used in jam for example), gives foods a firmer texture.

6.1.19 Sweeteners

Sweeteners are added into the food products to increase to stabilize the sweetness level. As we know that sugars are divided into different groups like fructose, sucrose, and maltose. Food industries also employed the use of artificial sweeteners like Aspartame for sweetening.

6.1. 20 Thickeners

Thickening agents can increase the thickness or viscosity of food products in a liquid or semi-liquid state. We can use CMC (Carboxymethyl Cellulose), Xanthan gum, and pectin (A polysaccharide) for this purpose.

ABOUT THE AUTHOR

A Co-Authored Book:

This book is written by Nouman Tariq who is a Food Engineer. He is a passionate Food Engineer with a superb work ethic. He has an interest in food sciences, physical properties of food materials, food processing technology, quality compliance, and audits (Process and Product) of food industries. He current aims are to write authentic books related to food sciences, food technology, food engineering, chemical engineering to help millions and trillions of food professionals and students to boost their career.

Muhammad Waseem Akbar is a food engineer by education. He is expert in nutritional, health and processing courses. He noticed that the bookish knowledge of diseases concerned with nutrition is present a lot in our society. Almost everyone has fatigue issues, stunting, stress problems and improper eating patterns lead towards life-ending diseases. So, our team want to educate society, so a genuine guide can help them to have good physical and mental health.

www.ingramcontent.com/pod-product-compliance
Lightning Source LLC
Chambersburg PA
CBHW071227130726
47998CB00002B/860